YOUR KNOWLEDGE HAS VALUE

- We will publish your bachelor's and
  master's thesis, essays and papers

- Your own eBook and book -
  sold worldwide in all relevant shops

- Earn money with each sale

Upload your text at www.GRIN.com
and publish for free

Internet of Things in Healthcare. Applications and Challenges

Hardik Modi
Shaili Joshi

Bibliographic information published by the German National Library:

The German National Library lists this publication in the National Bibliography; detailed bibliographic data are available on the Internet at http://dnb.dnb.de.

ISBN: 9783389022955
This book is also available as an ebook.

© GRIN Publishing GmbH
Trappentreustraße 1
80339 München

All rights reserved

Print and binding: Books on Demand GmbH, Norderstedt, Germany
Printed on acid-free paper from responsible sources.

The present work has been carefully prepared. Nevertheless, authors and publishers do not incur liability for the correctness of information, notes, links and advice as well as any printing errors.

GRIN web shop: https://www.grin.com/document/1472314

Exploring Applications and Challenges of IoT in Healthcare: A Comprehensive Review

Abstract:

This review paper provides an overview of the Internet of Things (IoT) in healthcare, focusing on its applications, and challenges. IoT has transformed healthcare through the introduction of remote patient monitoring, wearables, telemedicine, and the Internet of Medical Things (IoMT). These applications offer unprecedented opportunities for improving patient care, enhancing medical diagnostics, and increasing accessibility to healthcare services. Despite the numerous benefits, IoT in healthcare also faces several challenges. Overcoming these obstacles is critical to realising the full promise of IoT in healthcare and ensuring its safe and effective implementation. This review discusses strategies and technologies to overcome security and privacy issues, manage data overload, and improve internet connectivity, enabling a better connected more effective healthcare environment.

Keywords: Internet of Things(IoT), Healthcare, Remote Patient Monitoring, Sensor, Internet of Medical Things(IoMT)

Introduction

The way we interact with technology and data is being revolutionized by the Internet of Things (IoT), which has emerged as a revolutionary force across a number of sectors. IoT has created previously unthinkable opportunities for efficiency and creativity by allowing everyday devices to communicate with one another and share information by connecting them to the internet. IoT technologies are revolutionizing operations, optimizing resource utilization, and improving decision-making processes across a range of sectors. Industries may increase productivity, sustainability, and competitiveness in a world that is becoming more digitally connected by utilizing data analytics and connected devices. As IoT develops further, it has the potential to completely transform sectors and spur unprecedented levels of productivity, connectedness, and creativity. The incorporation of the Internet of Things (IoT) into numerous sectors provides great breakthroughs, particularly in healthcare. [1].

In the realm healthcare, IoT stands ready to transform the terrain of medical service provision, introducing a fresh era of tailored and interconnected healthcare solutions. This study examines the wide-ranging effects of IoT in healthcare, including its applications, challenges, and implications for patient welfare, healthcare professionals, and healthcare organizations. The impact of IoT extends beyond remote patient monitoring and sophisticated medical gadgets; it also includes instant access to data analytics has resulted in significant changes to the accessibility, delivery, and experience of health care services.

Remote patient monitoring tools enabled by IoT allow healthcare providers to monitor a patient's vital signs from a distance monitoring health markers from distant locations, permitting beforehand detection of health concerns and timely management. Additionally, smart medical devices like wearable health monitors and insulin pumps offer patients instant feedback and aid in managing chronic conditions while encouraging healthier lifestyle choices.

Moreover, healthcare organizations can enhance resource management, streamline processes, and enhance patient outcomes by leveraging insights derived from IoT-generated data. By analyzing extensive data sets collected from IoT devices and sensors,Medical experts can spot patterns, foresee health hazards, and tailor therapies to match the specific needs of every individual. This leads to improved quality of care and reduced healthcare costs. By merging IoT data with predictive analytics, potential health threats can be detected early [2].

The purpose of this review paper is to explore how the convergence of IoT and healthcare presents revolutionary opportunities for improving the provision of medical services. By delving into present applications, obstacles, and future prospects, it aims to stimulate additional research and innovation in utilizing IoT to enhance healthcare outcomes and elevate patient experiences.[1-45]

Application of IoT in Healthcare

The convergence of IoT and healthcare brings about a revolutionary transformation known as IoT-Enabled Healthcare. This pioneering integration merges cutting-edge IoT technology with healthcare services, presenting unparalleled possibilities for improving patient care, real-time health monitoring, and refining medical interventions[3][4]. IoT-Enhanced Healthcare represents a substantial shift in the healthcare industry, promoting a patient-centric, data-driven approach that enhances efficiency and connectivity. This transformative model not only places emphasis on patient needs but also utilizes data to optimize healthcare delivery and achieve better outcomes [4][5][6][7]. Several instances of IoT applications in the healthcare sector are elaborated upon below. Figure 1 depicts a few applications of IoT in the healthcare sector.

This image has been removed due to copyright issues.

Figure   1.   FEW APPLICATIONS OF IOT IN HEALTHCARE

Remote Patient Monitoring

The versatility of Internet of Things (IoT) technology provides patients with numerous opportunities to proactively engage in their healthcare. Key IoT applications for patients encompass various aspects of health management, including remote monitoring of vital signs, medication adherence support, and access to telemedicine services for remote consultations and diagnoses. Remote patient monitoring (RPM) can effectively prevent the progression to acute care and has the potential to enhance the management of chronic diseases.

The integration of IoT in healthcare marks the onset of a fresh era of ongoing remote monitoring of crucial health indicators such as blood pressure, body temperature, heart rate, and blood glucose levels [4]. Several kinds of sensors are used by the Internet of Things in healthcare to help with remote patient monitoring, such as Thermopile sensor, Photo-optic sensor, Image sensor, IR sensor**Error! Reference source not found.**. Studies show significant hospitalization reduction and nursing visit decrease with remote monitoring, challenging assumptions about managing severe clinical cases [5]. A study proposed an IoT solution using 5G to monitor cancer patients' body conditions during chemotherapy, by integrating condition measurement devices, predicting drug reactions to prevent adverse effects [9]. Recent research proposes automated Alzheimer's disease monitoring and distinguishes between healthy and unhealthy individuals[10][11][12][13]. Emerging technologies like deep learning are vital in medical IT. This system integrates IoT and deep learning for precise medication monitoring, improving accuracy and user experience remotely [15]. A different study proposed a real-time platform with a camera-equipped pillbox to detect and prevent interactions, enhancing adherence and enabling remote adjustments [16]. This paper [17] proposed a smart ambulance navigation system to address delays in sharing patient information. Sensors provide quick diagnosis, wirelessly transmitted to hospitals for prompt treatment. Patients can request ambulances and send SMS to emergency contacts, saving crucial time. The figure 2 illustrates the aspects monitored through remote or real-time monitoring.

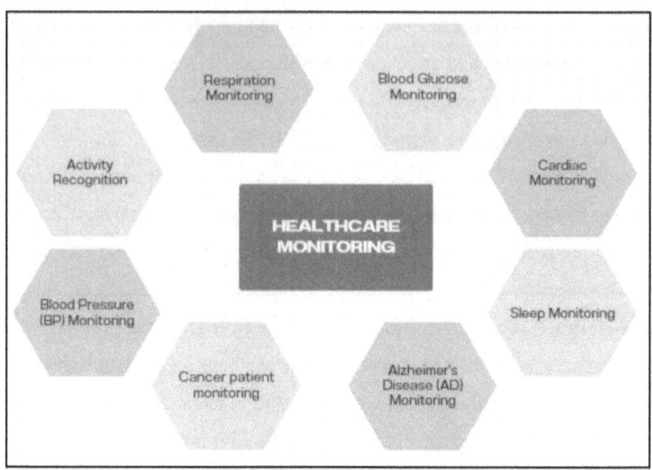

Figure 2. ASPECTS MONITORED THROUGH REMOTE/REAL-TIME MONITORING

Wearbale Technology

At the convergence of IoT and healthcare, wearable technology is significant because it offers inventive methods to record and promote individual health and wellness. Every IoT-based healthcare system comprises a sensor layer, gathering data from users/patients by measuring their vital signs and other essential signals, converting them into monitorable data. This layer employs various sensors, often wearable, allowing users to continue daily activities uninterrupted. Many systems utilize BAN or WSN for networking [18]. Wearable devices consist of a number of sensor nodes, each with a radio transceiver, low-speed computing unit, and limited memory. These sensors monitor parameters related to health such as SpO2, BP, temperature, EDA, ECG, HR, and RR [19]. Bluetooth, infrared, NFC, RFID, Wi-Fi, and Zigbee enable communication with smart phones and other devices [20].

These gadgets, worn directly on the body, gather live data and wirelessly transmit it to other devices or platforms for analysis and understanding. This paper [21] proposes an advanced AI-driven approach integrating with wearable Continuous Glucose Monitoring (CGM) devices for accurate glucose prediction and alert users to critical levels, crucial for type-1 Diabetes management. The frequency of Alzheimer's disease (AD) and Parkinson's disease (PD) highlights the urgent need for early detection of neurological symptoms. In [22], a prototype of sensor-equipped glasses capable of detecting Essential Tremor (ET) and counting Eye Blinks (EBs) without prior training. Experimental results demonstrate high accuracy in ET recognition and EB counting.

There are advancements in wearable biosensor technology, utilizing various mechanisms and materials from metals and semiconductors to flexible 2D materials, polymers, and bio-materials, have enabled the detection of analytes in bio fluids in ambient conditions. These biosensors have promise for both health care and monitoring the environment, as they interface with transmission of signals systems like Bluetooth, WiFi, and RF [23][24].

Telemedicine

The Internet of Things allows physicians and medical workers to accomplish jobs more efficiently and intelligently [25]. As an instance, monitoring can move from hospitals to homes, lowering resource demand and allowing elderly individuals to live independently for extended periods of time. It enables remote consultations, improves healthcare accessibility, and cuts travel time and expenses for both patients and doctors [26]. Telemedicine with IoT enables personalized healthcare through tailored treatment plans derived from ongoing data streams. In addition, telemedicine empowered by IoT has the capacity to enhance preventive care through the analysis of real-time data on health to identify warning signs and early symptoms of disease [10].

Scalability is crucial in telemedicine. Leading healthcare systems have expanded their services by using IoT devices to ensure continuous care delivery, particularly during public health emergencies. [27]. This transition to preventive care, that emphasizes on proactive efforts to maintain health and prevent disease before it occurs, has the potential to drastically lower healthcare expenditures while also improving population health overall [28].

Internet of Medical Things(IoMT)

The Internet of Medical Things (IoMT) is a rapidly emerging field within the broader Internet of Things (IoT), garnering increasing attention from researchers due to its wide-ranging applications in Smart Healthcare Systems (SHS) [29]. IoMT significantly enhances the accuracy, reliability, and productivity of electronic healthcare devices. This paper [30] provides an overview of IoMT, emphasizing enabling techniques such as RFID, AI, and blockchain in SHS. It presents a comparative analysis of IoMT architectures and explores various health domains, sensor applications, and protocol design challenges. The global pandemic emphasizes the importance of timely, on-site infectious disease testing. Point-of-care testing (POCT) provides rapid, real-time data that are critical for timely treatment and disease control. Integration with the Internet of Medical Things (IoMT) improves the connection and effectiveness of POCT devices, which are critical for managing infectious infections [31]. Whereas, this research [32] aims to employ machine learning classification algorithms for heart disease prediction. IoMT-based cloud-fog diagnostics are proposed, utilizing the fog layer for quick data analysis and simulation results show significant improvements. Another study [33] developed a real-time remote monitoring system for heart rate and body temperature using IoMT. It involves hardware and software design, sensor calibration, web server creation, and testing, with calibration results showing low error rates.

We can say that the Internet of Things has numerous implications for improving patient care and healthcare services. As it enables remote monitoring, allows for quicker interventions, and improves communication between patients and healthcare providers.

Challenges

The implementation of the Internet of Things (IoT) in healthcare has immense potential to transform medical practices, improve patient care, and enhance overall healthcare outcomes.However, the resulting integration poses various issues that demand scrutiny. From ensuring data confidentiality and safety to addressing interoperability issues, complying with regulations, ensuring reliable internet access, and overcoming user acceptance barriers are some of the challenges faced. Navigating these complexities requires careful examination and strategic planning. We will explore the primary challenges to adopting IoT in healthcare.

Data Security and Privacy

Data privacy is crucial given the ongoing collection of confidential patient medical information by IoT gadgets. Data breaches and unauthorized access pose significant threats, highlighting the significance of effective encryption and authentication, and access controls to protect patient information. Without sufficient security measures, or outdated software or firmware, patient data may be susceptible to hackers, potentially resulting in breaches of confidentiality and misuse of personal health information, jeopardizing patient trust and jeopardizing healthcare delivery. As a result, implementing strict security standards is critical to ensuring the confidentiality, integrity, and availability of patient information in IoT-powered medical systems.Furthermore, IoMT's reliance on networks that are interconnected renders it highly susceptible to network-based attacks such as man-in-the-middle attacks or lack of service situations, which jeopardise the integrity and accessibility of healthcare services [12]. A study [34] presents IDP, a data placement method tailored for IoT with privacy protection. The goal is to reduce data access time, utilisation of resources, and consumption of energy while maintaining privacy. This is accomplished through the application of the Non-dominated Sorting Genetic Algorithm II (NSGA-II) [35]. Another research [36] aims to develop a framework for data privacy using a biometric-based security system tailored for a restricted in resources wearable health tracking system.

Internet Connectivity

Internet connectivity is a crucial obstacle in implementing IoT in healthcare. Many healthcare facilities, particularly those located in distant or underdeveloped locations, may have limited or no dependable internet access. This can impede real-time data transfer from IoT devices, reducing the speed and accuracy of patient monitoring and actions.Furthermore, challenges with internet connectivity may restrict the interoperability of IoT devices and systems, preventing seamless communication and data sharing between healthcare practitioners and institutions. In cases where internet connectivity is unreliable, alternate options such as offline data storage or local network connections may be required, like addition of a hardware module GSM sim 900A [37], but they are not always practical or efficient. As a result, overcoming internet connectivity issues is critical to realizing the full potential of IoT in healthcare and guaranteeing equal access to quality medical services for all people, regardless of geographic location.

Data Prediction and Overload

Because of the volume of data created, the necessity for prompt diagnosis and decision-making, and the importance of responding quickly to observed abnormalities, uploading all data to the Cloud for analysis may be impractical. It is impractical to move massive amounts of data to the Cloud for analysis and storage, thereby delaying health-related decision-making processes. Current cloud computing systems are thought to be less capable of managing the vast amounts of data created by IoT [26].

Prediction problems arise in systems with various data-sets that can be perceived differently, complicating the fulfillment of defined goals. This factor is crucial in IoT and machine learning technology. Users require training to accurately forecast and understand data. Inaccurate interpretations may result in erroneous analysis, compromising healthcare service delivery. As a result, sufficient training is required to ensure the effective use of IoT data, improve healthcare outcomes, and reduce errors in medical decision-making [38][39].

Other Challenges

Interoperability is a hurdle with IoMT systems since data interchange between several IoMT-based systems is complex. Standardized interfaces are required to address this problem. Furthermore, integrating sensors and healthcare equipment may be costly for upgrading and upkeep. Including low-cost and free of servicing sensors can help to accelerate the growth and adoption of IoMT-based devices. Furthermore, the power usage of devices used in IoMT is a significant challenge, necessitating backup systems or large-capacity batteries. Sustainable medical gadgets are critical for tackling this issue. One of the issues posed by the growth of the Internet of Things (IoT) is integrating data from cyber-physical systems with existing traditional information systems [12][26].

Conclusion

In conclusion, the integration of IoT with health care has transformed the field by offering novel solutions to improve patient care and management. IoT technology has enabled healthcare practitioners to provide personalized, efficient, and remote treatment to patients via applications such as remote patient monitoring, wearables, telemedicine, and the Internet of Medical Things (IoMT), hence improving overall health outcomes and quality of life.

However, the implementation of IoT in healthcare poses considerable hurdles. Security and privacy concerns remain paramount, since IoT devices transmit and retain large volumes of sensitive patient data that are subject to assaults. To protect patient information, it is necessary to implement strong security measures and adhere to regulatory regulations. Furthermore, the proliferation of IoT devices has resulted in data overload, presenting healthcare providers with massive amounts of information. Effective data management is critical for generating valuable insights and actionable intelligence. Furthermore, establishing consistent internet connectivity in all healthcare settings is critical to the smooth operation of IoT devices and ongoing patient care. Despite these challenges, IoT's potential in healthcare is apparent. With continuing study, innovation, and collaboration among technology developers, healthcare providers, and regulatory agencies, the benefits of IoT can be maximized while the hazards are mitigated. By addressing security, privacy, data management, and connectivity challenges, the Internet of Things has the ability to improve healthcare delivery, making it more efficient, accessible, and patient-centered.

References

[1]   Ghozali, Muhammad Thesa. "Implementation of the IoT-based technology on patient medication adherence: A comprehensive bibliometric and systematic review." Journal of Information and Communication Technology 22, no. 4 (2023): 503-544.

[2]   Kazi Kutubuddin Sayyad Liyakat. "IoT in Healthcare". Medicon Medical Sciences 5.2 (2023): 01

[3]   Subramaniam, Shankar, Naveenkumar Raju, Abbas Ganesan, Nithyaprakash Rajavel, Maheswari Chenniappan, Chander Prakash, Alokesh Pramanik, Animesh Kumar Basak, and Saurav Dixit. "Artificial intelligence technologies for forecasting air pollution and human health: A narrative review." Sustainability 14, no. 16 (2022): 9951.

[4]   Vatin, Nikolai Ivanovich, Ruby Pant, Chandra Mohan, Anil Kumar, and N. Rajasekhar. "IoT-Enhanced Healthcare: A Patient Care Evaluation Using the IoT Healthcare Test." In BIO Web of Conferences, vol. 86, p. 01092. EDP Sciences, 2024.

[5]   Alshamrani, Mazin. "IoT and artificial intelligence implementations for remote healthcare monitoring systems: A survey." Journal of King Saud University-Computer and Information Sciences 34, no. 8 (2022): 4687-4701.

[6]   Othman, Soufiene Ben, Faris A. Almalki, Chinmay Chakraborty, and Hedi Sakli. "Privacy-preserving aware data aggregation for IoT-based healthcare with green computing technologies." Computers and Electrical Engineering 101 (2022): 108025.

[7]   SwornaNabila Sabrin, AKM Muzahidul Islam, Swakkhar Shatabda, and Salekul Islam. "Towards development of IoT-ML driven healthcare systems: A survey." Journal of Network and Computer Applications 196 (2021): 103244.

[8]   A.Vinitha IOT in healthcare: A Survey, Journal of Education Rabindra Bharati University, Volume XXIII, Issue 8, September 2021, ISSN: 0972 – 7175, (UGC CARE)

[9]   Khondokar, Tanvina, Saiful Islam, Tasmia Ishrat Alam Chadni, and Iftakhar Ali Khandokar. "A Drug Dose Control and Cancer Patient Monitoring System Based on the Internet of Things (IoT) with Integrated Cloud and 5G Protocol." In 2023 26th International Conference on Computer and Information Technology (ICCIT), pp. 1-6. IEEE, 2023.

[10]   Auwal, Aminu Muhammad. "IoT integration in telemedicine: Investigating the role of Internet of things devices in facilitating remote patient monitoring and data transmission." (2023).

[11]   Jethawa, Neha R., Pooja B. Gajghane, and Jayesh K. Kokate. "IOT BASED HEALTHCARE SYSTEM." International Journal of Creative Research Thoughts, Volume 6, Issue 1 March 2018, ISSN: 2320-2882

[12]   Umer, Muhammad, Turki Aljrees, Hanen Karamti, Abid Ishaq, Shtwai Alsubai, Marwan Omar, Ali Kashif Bashir, and Imran Ashraf. "Heart failure patients monitoring using IoT-based remote monitoring system." Scientific Reports 13, no. 1 (2023): 19213.

[13]   Khan, Afreen, and Swaleha Zubair. "An improved multi-modal based machine learning approach for the prognosis of Alzheimer's disease." Journal of King Saud University-Computer and Information Sciences 34, no. 6 (2022): 2688-2706.

[14]   Roh, Hyeji, Seulgi Shin, Jinseo Han, and Sangsoon Lim. "A deep learning-based medication behavior monitoring system." Math. Biosci. Eng 18, no. 2 (2021): 1513-1528.

[15]   Chavez, Eduardo, Billy Sifuentes, Ricardo Vidal, Juan Grados, Santiago Rubiños, and Abilio Cuzcano. "Remote monitoring applying IoT to improve control of medication adherence in geriatric patients with a complex treatment regimen, Lima-Peru." In Proceedings of the 2020 3rd international conference on electronics and electrical engineering technology, pp. 49-54. 2020.

[16]   Karagiannis, Dimitrios, Konstantinos Mitsis, and Konstantina S. Nikita. "Development of a low-power IoMT portable pillbox for medication adherence improvement and remote treatment adjustment." Sensors 22, no. 15 (2022): 5818.

[17]   Poongodi, Manoharan, Ashutosh Sharma, Mounir Hamdi, Ma Maode, and Naveen Chilamkurti. "Smart healthcare in smart cities: wireless patient monitoring system using IoT." The Journal of Supercomputing (2021): 1-26.

[18]   Mamdiwar, Shwetank Dattatraya, Zainab Shakruwala, Utkarsh Chadha, Kathiravan Srinivasan, and Chuan-Yu Chang. "Recent advances on IoT-assisted wearable sensor systems for healthcare monitoring." Biosensors 11, no. 10 (2021): 372.

[19]   Alekya, R., Neelima Devi Boddeti, K. Salomi Monica, R. Prabha, and V. Venkatesh. "IoT based smart healthcare monitoring systems: A literature review." Eur. J. Mol. Clin. Med 7 (2021): 2020.

[20]   Abdulmalek, Suliman, Abdul Nasir, Waheb A. Jabbar, Mukarram AM Almuhaya, Anupam Kumar Bairagi, Md Al-Masrur Khan, and Seong-Hoon Kee. "IoT-based healthcare-monitoring system towards improving quality of life: A review." In Healthcare, vol. 10, no. 10, p. 1993. MDPI, 2022.

[21]   Nasser, Ahmed R., Ahmed M. Hasan, Amjad J. Humaidi, Ahmed Alkhayyat, Laith Alzubaidi, Mohammed A. Fadhel, José Santamaría, and Ye Duan. "Iot and cloud computing in health-care: A new wearable device and cloud-based deep learning algorithm for monitoring of diabetes." Electronics 10, no. 21 (2021): 2719.

[22]   Sciarrone, Andrea, Igor Bisio, Chiara Garibotto, Fabio Lavagetto, Gerhard Staude, and Andreas Knopp. "A wearable prototype for neurological symptoms recognition." In ICC 2020-2020 IEEE International Conference on Communications (ICC), pp. 1-7. IEEE, 2020.

[23]   Wen, Feng, Tianyiyi He, Huicong Liu, Han-Yi Chen, Ting Zhang, and Chengkuo Lee. "Advances in chemical sensing technology for enabling the next-generation self-sustainable integrated wearable system in the IoT era." Nano Energy 78 (2020): 105155.

[24]   Verma, Damini, Kshitij RB Singh, Amit K. Yadav, Vanya Nayak, Jay Singh, Pratima R. Solanki, and Ravindra Pratap Singh. "Internet of things (IoT) in nano-integrated wearable biosensor devices for healthcare applications." Biosensors and Bioelectronics: X 11 (2022): 100153.

[25]   Aghdam, Zahra Nasiri, Amir Masoud Rahmani, and Mehdi Hosseinzadeh. "The role of the Internet of Things in healthcare: Future trends and challenges." Computer methods and programs in biomedicine 199 (2021): 105903.

[26]   Frimpong, Bismark Atta, Cláudia Barbosa, and Raed A. Abd-Alhameed. "The Impact of the Internet of Things (IoT) on Healthcare Delivery: A Systematic Literature Review." Journal of Techniques 5, no. 3 (2023).

[27]   Rahman, M., FM Javed Mehedi Shamrat, Mohammod Abul Kashem, Most Fahmida Akter, Sovon Chakraborty, Marzia Ahmed, and Shobnom Mustary. "Internet of things based electrocardiogram monitoring system using machine learning algorithm." Int. J. Electr. Comput. Eng 12, no. 4 (2022): 3739-3751.

[28]   Ferrier, Clément, Babak Khoshnood, Ferdinand Dhombres, Hanitra Randrianaivo, Isabelle Perthus, Jean-Marie Jouannic, and Isabelle Durand-Zaleski. "Cost and outcomes of the ultrasound screening program for birth defects over time: a population-based study in France." BMJ open 10, no. 7 (2020): e036566.

[29]   Joyia, Gulraiz J., Rao M. Liaqat, Aftab Farooq, and Saad Rehman. "Internet of medical things (IoMT): Applications, benefits and future challenges in healthcare domain." J. Commun. 12, no. 4 (2017): 240-247.

[30]   Srivastava, Jyoti, Sidheswar Routray, Sultan Ahmad, and Mohammad Maqbool Waris. "Internet of Medical Things (IoMT)-based smart healthcare system: Trends and progress." Computational Intelligence and Neuroscience 2022 (2022).

[31]   Jain, Shikha, Monika Nehra, Rajesh Kumar, Neeraj Dilbaghi, TonyY Hu, Sandeep Kumar, Ajeet Kaushik, and Chen-zhong Li. "Internet of medical things (IoMT)-integrated

biosensors for point-of-care testing of infectious diseases." Biosensors and Bioelectronics 179 (2021): 113074.

[32]   Chakraborty, Chinmay, and Amit Kishor. "Real-time cloud-based patient-centric monitoring using computational health systems." IEEE transactions on computational social systems 9, no. 6 (2022): 1613-1623.

[33]   Laila, Ida, A. Arifin, and Bidayatul Armynah. "Internet of medical things (iomt)-based heart rate and body temperature monitoring system." Indonesian Physical Review 5, no. 1 (2022): 1-14.

[34]   Xu, Xiaolong, Shucun Fu, Lianyong Qi, Xuyun Zhang, Qingxiang Liu, Qiang He, and Shancang Li. "An IoT-oriented data placement method with privacy preservation in cloud environment." Journal of Network and Computer Applications 124 (2018): 148-157.

[35]   Selvaraj, Sureshkumar, and Suresh Sundaravaradhan. "Challenges and opportunities in IoT healthcare systems: a systematic review." SN Applied Sciences 2, no. 1 (2020): 139.

[36]   Pirbhulal, Sandeep, Oluwarotimi Williams Samuel, Wanqing Wu, Arun Kumar Sangaiah, and Guanglin Li. "A joint resource-aware and medical data security framework for wearable healthcare systems." Future Generation Computer Systems 95 (2019): 382-391.

[37]   Sharma, Hitesh Kumar, J. C. Patni, Prashant Ahlawat, and Siddhratha Sankar Biswas. "Sensors based smart healthcare framework using internet of things (IoT)." Int J Sci Technol Res 9, no. 2 (2020): 1228-34.

[38]   Khan, Muhammad Adnan. "Challenges facing the application of IoT in medicine and healthcare." International Journal of Computations, Information and Manufacturing (IJCIM) 1, no. 1 (2021).

[39]   Bhuiyan, Mohammad Nuruzzaman, Md Mahbubur Rahman, Md Masum Billah, and Dipanita Saha. "Internet of things (IoT): A review of its enabling technologies in healthcare applications, standards protocols, security, and market opportunities." IEEE Internet of Things Journal 8, no. 13 (2021): 10474-10498.

[40]   Bhushan, Bharat, Avinash Kumar, Ambuj Kumar Agarwal, Amit Kumar, Pronaya Bhattacharya, and Arun Kumar. "Towards a secure and sustainable internet of medical things (iomt): Requirements, design challenges, security techniques, and future trends." Sustainability 15, no. 7 (2023): 6177.

[41]   Ajagbe, Sunday Adeola, Joseph Bamidele Awotunde, Ademola Olusola Adesina, Philip Achimugu, and T. Ananth Kumar. "Internet of medical things (IoMT): applications, challenges, and prospects in

a data-driven technology." Intelligent Healthcare: Infrastructure, Algorithms and Management (2022): 299-319.

[42] Sobin, C. C. "A survey on architecture, protocols and challenges in IoT." Wireless Personal Communications 112, no. 3 (2020): 1383-1429.

[43] Tariq, Noshina, Farrukh Aslam Khan, and Muhammad Asim. "Security challenges and requirements for smart internet of things applications: A comprehensive analysis." Procedia Computer Science 191 (2021): 425-430.

[44] Zeadally, Sherali, Farhan Siddiqui, Zubair Baig, and Ahmed Ibrahim. "Smart healthcare: Challenges and potential solutions using internet of things (IoT) and big data analytics." PSU research review 4, no. 2 (2020): 149-168.

[45] De Guzman, Keshia R., Centaine L. Snoswell, Monica L. Taylor, Leonard C. Gray, and Liam J. Caffery. "Economic evaluations of remote patient monitoring for chronic disease: a systematic review." Value in Health 25, no. 6 (2022): 897-913.

YOUR KNOWLEDGE HAS VALUE

- We will publish your bachelor's and
  master's thesis, essays and papers

- Your own eBook and book -
  sold worldwide in all relevant shops

- Earn money with each sale

Upload your text at www.GRIN.com
and publish for free